CW00602086

BRIDGES

The Francis Frith Collection

First published in the United Kingdom in 2004 by
Frith Book Company Ltd

ISBN 1-85937-900-1

British Library Cataloguing in Publication Data

Francis Frith's 50 Classics - Bridges
Compiled by Terence Sackett and Julia Skinner

Frith Book Company Ltd
Frith's Barn, Teffont,
Salisbury, Wiltshire SP3 5QP
Tel: +44 (0) 1722 716 376
Email: info@francisfrith.co.uk
www.francisfrith.co.uk

Printed and bound in Great Britain

Front Cover: MENAI, The Suspension Bridge 1890 23187
The colour-tinting is for illustrative purposes only, and is not intended to be historically accurate.

FRANCIS FRITH
VICTORIAN PIONEER

Francis Frith, founder of the world-famous photographic archive, was a complex and multi-talented man. A devout Quaker and a highly successful Victorian businessman, he was philosophic by nature and pioneering in outlook. By 1855 he had already established a wholesale grocery business in Liverpool, and sold it for the astonishing sum of over £200,000, which is the equivalent today of over £15,000,000. Now in his thirties, and captivated by the new science of photography, Frith set out on a series of pioneering journeys up the Nile and to the Near East.

INTRIGUE AND EXPLORATION

He was the first photographer to venture beyond the sixth cataract of the Nile. Africa was still the mysterious 'Dark Continent', and Stanley and Livingstone's historic meeting was a decade into the future. The conditions for picture taking confound belief. He laboured for hours in his wicker dark-room in the sweltering heat of the desert, while the volatile chemicals fizzed dangerously in their trays. Back in London he exhibited his photographs and was 'rapturously cheered' by members of the Royal Society. His reputation as a photographer was made overnight.

VENTURE OF A LIFE-TIME

By the 1870s the railways had threaded their way across the country, and Bank Holidays and half-day Saturdays had been made obligatory by Act of Parliament. All of a sudden the working man and his family were able to enjoy days out, take holidays, and see a little more of the world.

With typical business acumen, Francis Frith foresaw that these new tourists would enjoy having souvenirs to commemorate their days out. For the next thirty years he travelled the country by train and by pony and trap, producing fine photographs of seaside resorts and beauty spots that were keenly bought by millions of Victorians.

These prints were painstakingly pasted into family albums and pored over during the dark nights of winter, rekindling precious memories of summer excursions. Frith's studio was soon supplying retail shops all over the country, and by 1890 F Frith & Co had become the greatest specialist photographic publishing company in the world, with over 2,000 sales outlets, and pioneered the picture postcard.

FRANCIS FRITH'S LEGACY

Francis Frith had died in 1898 at his villa in Cannes, his great project still growing. The archive he created continued in business for another seventy years. By 1970 it contained over a third of a million pictures showing 7,000 British towns and villages.

Frith's legacy to us today is of immense significance and value, for the magnificent archive of evocative photographs he created provides a unique record of change in the cities, towns and villages throughout Britain over a century and

more. Frith and his fellow studio photographers revisited locations many times down the years to update their views, compiling for us an enthralling and colourful pageant of British life and character.

We are fortunate that Frith was dedicated to recording the minutiae of everyday life. For it is this sheer wealth of visual data, the painstaking chronicle of changes in dress, transport, street layouts, buildings, housing, engineering and landscape that captivates us so much today, offering us a powerful link with the past and with the lives of our ancestors.

Computers have now made it possible for Frith's many thousands of images to be accessed almost instantly. The archive offers every one of us an opportunity to examine the places where we and our families have lived and worked down the years. Its images, depicting our shared past, are now bringing pleasure and enlightenment to millions around the world a century and more after his death.

INTRODUCTION

This attractive pocket book contains 50 classic Frith period photographs of bridges from all over England and Wales.

You'll find every type of bridge here, spanning modest streams, rivers, and broad estuaries. They include footbridges, clapper bridges, pack-horse bridges, medieval arched bridges, toll bridges, decorative Palladian bridges, suspension bridges, bascule bridges, canal bridges, and railway bridges. Each photograph is accompanied by an informative caption.

Evocative and atmospheric, these stunning images show British engineering at its most innovative and graceful.

This much-photographed packhorse bridge, allowing wool-laden mules, donkeys or horses to traverse the River Aller, is said to be from the 18th century. Presumably the barrel in the water under the bridge came from the former Pack Horse Inn, from which the photograph was taken.

The Grand Junction Canal, now the Grand Union, was built between 1793 and 1805 to link Birmingham with the River Thames and London. The Aylesbury Arm was a branch which ran from near Marsworth across the Vale to Aylesbury, and opened in 1815; an immediate consequence was the halving of coal prices in Aylesbury! This view looks along the towpath to the Park Street bridge.

AYLESBURY, THE CANAL 1897 39642

7

The medieval Kentish ragstone bridge at Aylesbury is on the site of an ancient ford, once the only crossing between Rochester and Maidstone. The bridge is perfectly proportioned, with one wide central span and three smaller arches on the approach.

This famous bridge spans the Mawddach estuary. The railway was built as part of the Cambrian railway, with two stations, Barmouth and Barmouth Junction. The bridge was opened in 1867 and its original design, as shown in this picture, included rolling sections that could be opened for river traffic to sail through. It is 800 yards long and also has a road for foot passengers.

Walney Bridge links Barrow-in-Furness with Walney island and Vickerstown, which was specially built to house the shipyard workers more than 100 years ago. Opened in 1908, it was originally a toll bridge. It was renamed the Jubilee Bridge in 1935. This photograph shows ship-builders swarming across the bridge from the dockyards at the end of a working day. Protected by the enclosing reef of Walney Island, Barrow flourished as a shipbuilding centre in the 19th and 20th centuries.

BARROW-IN-FURNESS, WALNEY BRIDGE 1912 64407

The village of Beddgelert has grown up in the mountains at the confluence of the Colwyn and Glaslyn rivers. A gracious two-span stone bridge spans the River Colwyn; the bridge was destroyed by flash floods in 1799 and again in 1906.

Bedford's medieval stone bridge was replaced by the Improvement Commissioners set up by Act of Parliament in 1803. The present bridge was designed by the local architect John Wing. Its foundation stone was laid by the Marquess of Tavistock, the eldest son of the Duke of Bedford, in 1811. The costs proved high. By the time the bridge opened in 1813, it was done without ceremony: the local MP, Samuel Whitbread, merely walked across to meet the Commissioners and shake hands. A further plaque records that it was opened free of tolls in 1835 - the debt by then had been paid off.

Bideford's twenty-four arched bridge has spanned the Torridge for seven hundred years. The span of the arches varies between 12 and 25 feet. Bideford, as its name suggests, stands near to a more ancient crossing point on the Torridge - the place 'by-the-ford'.

BIDEFORD, THE BRIDGE 1890 24792

13

Noted for its irregular arches, the bridge dates mainly from the 15th century, but some parts are thought to be much older.

A pack-horse bridge was originally built across the river at Bradford-on-Avon in the 13th century. The bridge has been repaired and altered over the years, and is now twice its original width; however some traces of the medieval bridge still survive on the eastern side of the river. The small building on the bridge was first a chapel and later used as a 'lock up,' in which the preacher John Wesley once spent an uncomfortable night.

BRIDGNORTH, THE BRIDGE 1898 42624

The first bridge, the one that gave the town its name, was built long ago. This bridge, however, has been described as a 'mongrel structure' because it is thought to incorporate parts of an earlier 14th-century structure within it. The earlier bridge had a chapel on it where travellers could stop and pray for a safe journey as they set off.

◀ Copied by James Gibbs from the one at Wilton House in Wiltshire, this bridge from 1738 crosses the end of the Octagon Lake. It is based on a design by the great Italian architect Andrea Palladio.

▶ Today the speed limit within Bristol's Floating Harbour is 6mph, and craft proceeding under the Prince Street Swing Bridge, Redcliffe Bascule Bridge or Bristol Bridge should sound one prolonged blast on their horn before doing so. Redcliffe Bascule Bridge is only raised by prior arrangement with the Harbour Master, but few craft need to worry about that as the clearance beneath the bridge is 3.6 metres.

BRISTOL, THE NEW
BASCULE BRIDGE
C1950   B212243

19

Wren's bridge, built by Robert Rumbold in 1709-12, has a balustraded parapet and heraldic beasts on display. It is also known as Kitchen Bridge; it seems that the master and fellows of St John's defied the architect and had it put at the end of the lane leading to the college kitchens.

CAMBRIDGE, ST JOHN'S COLLEGE 1890   26443

At the tiny village of Cark the River Eea flows under a low bridge into the sands of Morecambe Bay.

'Capability' Brown carried out substantial alterations to the grounds of Chatsworth House; the River Derwent was widened and diverted as part of his design. The three-arched bridge over the river was built by James Paine between 1758-64, purportedly from a design by Michael Angelo. It is often claimed that Chatsworth was Jane Austen's inspiration for Mr Darcy's home of Pemberley in 'Pride and Prejudice', following her visit there in 1811.

In 1852 a suspension bridge was built over the Dee to link the suburb of Queen's Park with the Groves on the north side of the river. The original suspension bridge was fairly narrow and unable to take any wheeled vehicle except for invalid carriage and perambulators, and was rebuilt in 1923.

In 1752, William Vick bequeathed money towards the eventual bridging of the Clifton Gorge. It was not until 1829 that a competition was held for engineers and architects to submit a design. Despite entries from the likes of Thomas Telford, the competition was won by the still relatively inexperienced Isambard Kingdom Brunel. Work halted when money ran out in 1843, and the bridge was not finally completed until 1864. The chains used to support the road were bought in second-hand, having been used on the old Hungerford Bridge in London.

Conwy's bridge was built by Telford in 1826, and measures 327 feet long. It was built for his Holyhead road, and replaced a notorious ferry across the dangerous waters of the Conwy estuary. It hangs on eight chains in two sets over two piers, with adjustment at one end into the rock under the castle, and at the other end into solid rock. This photograph also shows a tower of Stephenson's tubular railway bridge (centre of picture), which was built on the line of the old L & NWR Railway in 1848. A 19th century guide describes how it increases in height above high water from around 22 feet at the ends to 25 feet at the centre. The two tubes are each 14 feet wide and weigh around 1,300 tons.

COUNTESS WEAR, THE BRIDGE OVER THE EXE

◄ *Built in 1770, this bridge was (and is) the lowest crossing point on the River Exe (if you exclude the M5 motorway bridge, which is not a river crossing in the accepted sense). Note the fishermen with their nets. There used to be salmon in this river; today few are caught.*

▶ *Crowland grew up on a low island amid the surrounding marshes. The town is famous for its 14th-century triangular bridge, which had three footways over the beds of three streams. The streets no longer have streams flowing down them, so the bridge is now a redundant curiosity.*

CROWLAND,
THE BRIDGE 1894 34833

27

The tiny hump-backed bridge at Ashness on the narrow road which leads up from the eastern shore of Derwent Water to the Norse hamlet of Watendlath has been seen on countless Lake District calendars. The bold profile of Skaddaw fills the background across the lake.

DERWENT WATER, ASHNESS BRIDGE 1893   32870

This view shows six of the bridge's ten arches, four of which are on dry land. In medieval times a number of buildings were built on this bridge, including the chapels of St Andrew and St James which stood at either end. The latter was replaced by the House of Correction in 1632.

◀ *The bridge at Elstead has five stone arches which are 500 years old, and parts may be 200 years older still; however, the brick parapets were added about a hundred years ago.*

▶ *This famed packhorse bridge across the River Barle has been here for centuries; it is 180ft long with seventeen spans. To the right, the river is quite shallow normally, so that wheeled vehicles and horses not happy on the slabs can ford the river. The Barle can flood violently, and the bridge was washed away in 1947, 1950 and 1980. The slabs are now numbered so they can be retrieved and replaced in the correct order.*

EXMOOR,
TARR STEPS 1929   82159A

31

FLATFORD, BRIDGE COTTAGE 1907 57552

It was Flatford, and nearby East Bergholt, which provided the young John Constable with the inspiration for many of his paintings. Flatford Mill featured in several of his works. This thatched cottage is called, appropriately, Bridge Cottage.

Haven Bridge joins Great Yarmouth (on the far bank) with Southtown. The centre section lifts up to allow the passage of boats. It was replaced in 1930 at a cost of £200,000.

This wooden footbridge was built and paid for by public subscription in the year this photograph was taken. The structure, however, eventually rotted over the course of time; it has now been replaced by a stone bridge.

GUILDFORD, THE RIVER WAY,
NEW FOOTBRIDGE 1909   61967

35

The ford that gave the city of Hereford its name was probably a little downstream from here, below the site of the castle. The first bridge on this site was built in around 1125, but this stone bridge replaced it in 1490. For the next 200 years it was the only bridge crossing the River Wye in all of Herefordshire.

HEREFORD, THE CATHEDRAL AND WYE BRIDGE
1890   26957

The medieval bridge over the River Ouse was started in 1332 to connect Huntingdon with Godmanchester, and the respective authorities paid for three arches - resulting in different styles - with the builders starting on each bank and meeting in the middle.

◄ *Building the world's first iron bridge was an expensive venture for Abraham Darby III. The idea was first proposed by the Shropshire architect, Thomas Farnolls Pritchard, and a consortium was founded to finance it. In fact, when he died Darby was still paying off the debts that he had accrued from building it. A toll bridge, it opened to the public on 1 January 1781.*

▶ *Kirkby Lonsdale's famous Devil's Bridge over the River Lune traditionally gets its name because it was built by the Devil, who claimed the soul of the first being to cross it. In the event, it was nothing more than an old dog. In fact, the elegant, soaring structure was probably built in the 12th century.*

KIRKBY LONSDALE,
THE DEVIL'S BRIDGE 1899 42875

The cast iron bridge was built in the 1850s with Sir Charles Barry, the Houses of Parliament architect, acting as consultant.

Commuters flood across the bridge in the morning from London Bridge Station to their City offices. The hay cart is a reminder that horses needed fuel, and many stables were in Southwark, on the south bank. This bridge was rebuilt in Arizona after being replaced in 1972.

This important mid 15th-century bridge at the lowest crossing point of the River Fowey links the two parts of Lostwithiel. This is also the highest tidal point up to which shipping once came; but as this view shows, the quays and channels became silted by material washed down the river.

Boulter's Lock is probably the most famous lock on the Thames, and was the first and lowest on the river of the first set of eight to be built under the legislation on 1770. The lock has always been a popular spot for 'messing about' on boats. A boulter was another name for a miller.

MAIDENHEAD,
BOULTER'S LOCK BRIDGE 1906 54082

Maidenhead Bridge was designed by Sir Robert Taylor and rebuilt in 1777. At the height of the coaching era, up to five hundred coaches crossed the bridge every day.

44

MAIDENHEAD, THE BRIDGE 1906 54099

Thomas Telford's soaring Menai Suspension Bridge, 100ft (30m) above the straits between the mainland and Anglesey, was completed in 1826; it was the first structure of its type to carry heavy traffic. The pedestal towers were designed in the then fashionable Egyptian style. The bridge is 1,000ft (350m) long with a central span of 539ft (164m). The strait was bridged at the highest point because of the Admiralty's requirements that a fully rigged ship should be able to pass beneath.

◀ *Middlesbrough's bridge was opened in 1901. Pedestrian's and vehicles cross by means of a suspended platform which moves to and fro across the Tees. Whether it is worth the cost of the upkeep has long been a matter of debate. The bridge is often closed for repair, or because high winds make it dangerous to use. On the other hand, it is a symbol of Middlebrough's industrial past.*

▶ *The gate is a Norman structure dating from 1262, and it is a rare example of a fortified gateway on a bridge.*

MONMOUTH,
THE BRIDGE OVER THE
MONNOW 1891 28782

NEWBY BRIDGE, THE SWAN HOTEL

A turnpike road to Kendall once went through the village of Newby Bridge. A wooden bridge originally crossed the River Leven but was replaced with a five-arched stone bridge, built in 1651.

Northwich's bridge was originally built of stone, and was an early crossing point over the River Weaver. In medieval times the bridge was often thrown down by flood waters, and a regular ferry service conveyed people across the river. This photo shows Town Bridge just one year after it was constructed in 1899, looking west towards Winnington Street. The gentleman on the left of the picture wearing a peaked cap was the operator of the swing bridge. Town Bridge and Hayhurst Bridge (also in Northwich) are believed to be the first electrically powered swing bridges in Britain, and the first to be built on floating pontoons.

Dartmoor's clapper bridges, despite their pre-historic look, are actually medieval; they were constructed for the pack-horse trains that were the transport system of the moor. The giant slabs that make up the spans can weigh up to eight tons. The clapper bridges had to be strong, as heavy rain on the moor can cause the river to rise by three or four feet and change the river from a placid brook to an impassable torrent.

POSTBRIDGE, THE CLAPPER BRIDGE 1907 5788

The Royal Albert Bridge at Saltash was opened by Prince Albert in May 1859. Brunel's celebrated masterpiece across the Tamar estuary made the first direct rail link between Cornwall and the rest of England.

SELBY, THE OLD TOLL BRIDGE 1918 68170

The wooden toll-bridge over the Ouse was built in the 18th century.

Sonning's redbrick 18th-century bridge crosses the Thames next to the White Hart Hotel. The earliest part of the hotel dates back to the Elizabethan era, when it was a hostelry for those passing by on the river.

A small footbridge crosses the fast flowing stream that runs through Staithes. A young James Cook was an apprentice grocer in the village before the lure of the sea took him around the world on voyages of discovery.

STAITHES, THE BRIDGE 1886   18208

55

According to tradition, Tewkesbury's bridge was built by order of King John in 1197. Until the 19th century it was called the Long Bridge; it had undergone extensive repairs in the reign of Charles II.

Linking the east and west sides of the town over the River Esk, the bridge is a favourite place where people could stop for a chat or stand and watch the world go by. The small hut is the control point for the bridge, which could be raised to allow shipping through.

Built in the 12th century, the original bridge was of wood, and was probably sited nearer the Guildhall. Shops and houses, as many as fifty, were built on it. In 1565 the bridge collapsed, and the replacement structure was built of stone. The present bridge was built in the early 19th century.

INDEX

ALLERFORD, The Bridge 1900 45704	6
AYLESBURY, The Canal 1897 39642	7
AYLESFORD, The River Medway 1898 41549	8
BARMOUTH, The Railway Bridge 1896 37685	9
BARROW-IN-FURNESS, Walney Bridge 1912 64407	10
BEDDGELERT, The Bridge 1889 21832	11
BEDFORD, The Bridge 1921 70434	12
BIDEFORD, The Bridge 1890 24792	13
BIDFORD-ON-AVON, The Bridge 1910 62637	14
BRADFORD-ON-AVON 1900 45371	15
BRIDGNORTH, The Bridge 1898 42624	16–17
BRISTOL, The New Bascule Bridge c1950 B212243	19
BUCKINGHAM, The Palladian Bridge, Stowe c1960 B280068	18
CAMBRIDGE, St John's College 1890 26443	20
CARK, The Bridge 1897 40515	21
CHATSWORTH HOUSE and the Bridge 1886 18642	22
CHESTER, The Suspension Bridge 1914 67546	23
CLIFTON, The Bridge 1887 20167	24
CONWY, The Castle and Suspension Bridge 1906 54810	25
COUNTESS WEAR, The Bridge over the Exe 1906 53981	26
CROWLAND, The Bridge 1894 34833	27
DERWENT WATER, Ashness Bridge 1893 32870	28
DURHAM, Elvet Bridge 1918 68236	29
ELSTEAD, The Bridge 1906 53584	30
EXMOOR, Tarr Steps 1929 82159A	31
FLATFORD, Bridge Cottage 1907 57552	32–33
GREAT YARMOUTH, Haven Bridge 1896 37953	34
GUILDFORD, The River Way, New Footbridge 1909 61967	35
HEREFORD, The Cathedral and Wye Bridge 1890 26957	36
HUNTINGDON, The Old Bridge 1929 81872	37
IRONBRIDGE, The Bridge from the River 1892 30891	38
KIRKBY LONSDALE, The Devil's Bridge 1899 42875	39
LONDON, London Bridge c1880 L1303429	41
LONDON, Westminster Bridge, Queen Victoria Jubilee Day 1897 L130219	40
LOSTWITHIEL, The Bridge 1891 29844	42
MAIDENHEAD, Boulter's Lock Bridge 1906 54082	43
MAIDENHEAD, The Bridge 1906 54099	44
MENAI, The Suspension Bridge 1890 23187	45
MIDDLESBROUGH, The Transporter Bridge 1913 66412	46
MONMOUTH, The Bridge over the Monnow 1891 28782	47
NEWBY BRIDGE, The Swan Hotel 1914 67414	48
NORTHWICH, The Swing Bridge 1900 45422	49
POSTBRIDGE, The Clapper Bridge 1907 5788	50
SALTASH, The Ferry 1924 76023	51
SELBY, The Old Toll Bridge 1918 68170	52–53
SONNING, The Bridge 1904 52035	54
STAITHES, The Bridge 1886 18208	55
TEWKESBURY, King John's Bridge 1907 57679	56
WHITBY, The Bridge 1913 66266	57
YORK, The Ouse Bridge 1885 18463	58

FREE PRINT OF YOUR CHOICE

Choose any Frith photograph in this book.

Simply complete the Voucher opposite and return it with your remittance for £2.25 (to cover postage and handling) and we will print the photograph of your choice in SEPIA (size 11 x 8 inches) and supply it in a cream mount with a burgundy rule line (overall size 14 x 11inches).

Please note: photographs with a reference number starting with a "Z" are not Frith photographs and cannot be supplied under this offer.

Offer valid for delivery to UK addresses only.

PLUS: **Order additional Mounted Prints at HALF PRICE - £7.49 each** (normally £14.99) If you would like to order more Frith prints from this book, possibly as gifts for friends and family, you can buy them at half price (with no additional postage and handling costs).

PLUS: **Have your Mounted Prints framed**

For an extra £14.95 per print you can have your mounted print(s) framed in an elegant polished wood and gilt moulding, overall size 16 x 13inches (no additional postage and handling required).

FRITH PRODUCTS AND SERVICES

All Frith photographs are available for you to buy as framed or mounted prints. From time to time, other illustrated items such as Address Books, Calendars, Table Mats are also available. Already, almost 50,000 Frith archive photographs can be viewed and purchased on the internet through the Frith website.

For more detailed information on Frith companies and products, visit:

www.francisfrith.co.uk

Mounted Print
Overall size 14 x 11 inches (355 x 280mm)

IMPORTANT!

These special prices are only available if you use this form to order. You must use the ORIGINAL VOUCHER on this page (no copies permitted).

We can only despatch to one address. This offer cannot be combined with any other offer.

For further information, contact:

The Francis Frith Collection, Frith's Barn,

Teffont, Salisbury SP3 5QP

Tel: +44 (0) 1722 716 376

Fax: +44 (0) 1722 716 881

Email: sales@francisfrith.co.uk

Voucher

for FREE
and Reduced Price
Frith Prints

Do not photocopy this voucher. Only the original is valid, so please fill it in, cut it out and return it to us with your order.

Picture ref no	Page number	Qty	Mounted @ £7.49 UK	Framed + £14.95	UK orders Total £
1		1	Free of charge*	£	£
2			£7.49	£	£
3			£7.49	£	£
4			£7.49	£	£
5			£7.49	£	£
6			£7.49	£	£

Please allow 28 days for delivery

* Post & handling (UK)	£2.25
Total Order Cost	£

Title of this book . . . .

I enclose a cheque / postal order for £ . . . .
payable to 'The Francis Frith Collection'

OR debit my Mastercard / Visa / Switch (Maestro) / Amex card
(credit cards please on all overseas orders), details below

Card Number

Issue No (Switch only)   Valid from (Amex/Switch)

Expires   Signature

Name Mr/Mrs/Ms . . . .

Address . . . .

. . . .

. . . . Postcode . . . .

Daytime Tel No . . . .

E-mail . . . .

Valid to 31/12/07